Photographic reproduction of the original covers, completely unmarked save for the stamped numeral "1".
NAA: A3269, Q3

THE ART OF GUERILLA WARFARE

G.S.(R), THE WAR OFFICE

MAY, 1939

ISBN-13: 978-1976334627
ISBN-10: 1976334624

(©) C.A. Brown, 2017
All Rights Reserved

Cover Illustration: Extracted from Special Operations Executive Special Training School Headquarters Training Instruction No. 10 *"Illustrations of Firing Positions".*

G.S.(R)

CONTENTS

Introductory	vii
Object	1
Objectives of Guerilla Warfare	4
Methods and Principles	4
Organization	6
"The Chief"	9
Arms and Equipment	10
Information and Intelligence Services	11
Intercommunication	13
Training	14
Enemy Counter Action	15
Planning and Action	17
Preparatory Planning	18
Friendly Population	18
Hostile Population	19
Neutral Countries	20
Geographical	20
Organization of Bands	21
Conclusion	22

G.S.(R)

Introductory.

General Service (Research), or *G.S.(R)*, was the rather bland designation for what would later evolve into the "dirty tricks" department of the British War Office. Established in 1936, G.S.(R) was, by January 1939, tasked with development of various special operations, military intelligence, special weapons and equipment, and irregular warfare concepts in the years immediately preceding the outbreak of the Second World War.

Redesignated *Military Intelligence (Research)*, or *M.I.(R)*, in the Spring of 1939, with the crisis in Europe rapidly coming to a head, the organisation's work took on far greater urgency. M.I.(R) was expanded and became responsible for not only covert intelligence and irregular warfare, but also the development of novel anti-tank weapons and tactics in an attempt to blunt the sharp edge of Hitler's state-of-the-art armoured forces which, UK War Office planners had correctly determined, could very well ride roughshod across Europe if left unchecked.

Joining G.S.(R) in April 1939 was Major Colin Gubbins. Gubbins was an artillery officer who had seen service in the Great War in France, and significantly, had been involved in operations against guerrilla forces during the Russian Civil War in 1918/1919 and in Ireland against the IRA from 1919 to 1921. Later, while posted to the Northwestern Frontier in India, he had become familiar with irregular tribal warfare the use of human intelligence assets. He was perfectly placed to author three handbooks for irregular warfare - *The Art of Guerilla Warfare*, *The Partisan Leader's Handbook* and *How to Use High Explosives*.

Gubbins had the three handbooks completed by May of 1939, and then was dispatched to Poland to investigate the establishment of a "stay-behind" guerrilla force in the likely event of a German invasion of that country. As it stood, the lightning speed of the German invasion overtook the planning of the Polish "stay behind" force and Gubbins returned to the UK.

With the outbreak of war, Gubbins was tasked with establishing the UK's Independent Companies (later designated "Commandos"), which he joined in operations in Norway. Later still, Gubbins was tasked with the establishment of a "Stay-Behind" guerrilla force in the UK in advance of a threatened Nazi invasion of Britain.

THE ART OF GUERILLA WARFARE

This force was given the rather nondescript designation of "Auxiliaries" and were nominally placed on the war establishment of the Local Defence Volunteers (Home Guard). The Auxiliaries were formed into local bands called Patrols and were trained in commando tactics and supplied with a variety of weapons and demolitions stores which were cached around the countryside. Operating from underground "operational base" dugouts, patrols would harry and hinder occupying German forces while gathering intelligence which would be transmitted to the British government in exile in Canada via specially trained covert wireless units.

Gubbins was posted to the newly formed Special Operations Executive toward the end of 1940. G.S.(R) had morphed into M.I.(R) and in 1940 was merged with Section D of the Secret Intelligence Service to become the Special Operations Executive (SOE). An unusual organisation, SOE combined intelligence functions and special warfare, making it a self-contained covert force which later inspired the creation of the US Office of Strategic Services (OSS), which later itself inspired the creation of the United States Central Intelligence Agency (CIA). In 1943, Gubbins was appointed head of the SOE and oversaw operations all over the world, from Europe to North Africa to South East Asia and the China/Burma/India theatre.

The Art of Guerilla Warfare was a short training pamphlet which informed and influenced the guerrilla warfare training programs for the British Independent Companies (Commandos), the Home Guard "Stay Behind" Auxiliaries, the Special Operations Executive, resistance groups in Europe, the Australian and New Zealand Independent Companies, the Australian Services Reconnaissance Department (Z Special Unit), the Allied Intelligence Bureau's Philippines Regional Section and finally , the US Office of Strategic Services itself.

The pamphlet details the organisation and conduct of guerrilla warfare in occupied territories, from the establishment of a guerrilla band to intelligence collection and dissemination. It was influenced not only by Gubbins' own experiences in Russia, Ireland and India, but also by his intensive study of guerrilla warfare in general, from the Boer commandos in South Africa during the Boer War, to the guerrilla operations in the Spanish Civil War and the ongoing Chinese Civil War and Sino-Japanese war, where Communist leader Mao Tse Tung was using guerrilla tactics to fight both the Chinese Nationalists and the invading Japanese.

The Art of Guerilla Warfare and its companion volume, *The Partisan Leader's Handbook* are unique in that

G.S.(R)

they represent the first official British military doctrine on the conduct of guerrilla warfare against an occupying force. This they do in a brief and to the point manner, and if one reads closely, it becomes clear that there is still much wisdom contained within which could be of use to insurgent and counterinsurgent alike in the modern era.

CA Brown
July 2017

G.S.(R)

THE ART OF GUERILLA WARFARE.

General Principles.

Object

1. The object of guerilla warfare is to harass the enemy in every way possible within all the territory he holds to such an extent that he is eventually incapable either of embarking on a war, or of continuing one that may already have commenced. The sphere of action should include his home country, and also, in certain circumstances, such neutral countries as he uses as a source of supply. This object is achieved by compelling the enemy to disperse his forces in order to guard his flanks, his communications, his detachments, supply depots, etc., against the attacks of guerillas, and thus so to weaken his main armies that the conduct of a campaign becomes impossible.

2. There are three main types of guerilla warfare :—

  (a) The activities of individuals, or of small groups working by stealth on acts of sabotage.
  (b) The action of larger groups working as a band under a nominated leader, and employing military tactics, weapons etc., to assist in the achievement of their object, which is usually of a destructive nature.
  (c) The operations of large guerilla forces, whose strength necessitates a certain degree of military organization in order to secure their cohesion and to make and carry out effectively a plan of campaign.

3. The type of guerilla warfare, that can be carried out in any particular territory is dependent on the local conditions at the time, as explained later. The greater, however, should always include the less — i.e., where circumstances are favourable to the employment of large guerilla forces, they will also permit the action of partisan bands and of saboteurs; Where conditions are unsuitable to large scale operations, the action of partisan bands should be supported by that of saboteurs.

4. The culminating stage of guerilla warfare should always be to produce in the field large formations of guerillas, well-armed and well-trained, which are able to take a direct part in the fighting by attacks on suitable hostile formations and objects in direct

conjunction with the operations of the regular troops. It may well be, however, that, in the early days of the war, guerilla activities must, owing to the enemy's strength and to lack of support of the local population, be limited to acts of sabotage.

As the war progresses, and as the enemy's hold begins to weaken owing to successful sabotage, to war weariness of the enemy's troops, and as the inhabitants cease to be overawed, conditions will become ripe for the formation of partisan bands.

These bands will, at the commencement, act singly or in small local concentrations. By their audacity and apparent immunity from hostile counter-measures, they must then fan the flame of revolt until circumstances become favourable for the organization of large groups of bands, working under central leadership on a semi-military basis, necessitating a considerable degree of co-ordination as regards arrangements for supplies, munitions, collection of military intelligence, etc.

5. There are two main points in this connection to bear in mind:-

    (a) To obtain the maximum effect from guerilla warfare it is necessary to make use of all three types. Therefore, a careful study must be made as early as possible of the territories concerned, so as to determine for what methods of warfare each territory is suited, and to make the necessary preparations in advance. It is an extravagant waste of effort and opportunity if, for example, in an area suited for large scale guerilla operations, activities are, for want of preparation and forethought, limited to the uncoordinated actions of partisan bands and saboteurs.

    Further, it must be remembered that the enemy will institute counter-measures as soon as guerilla activities against him commence. If these activities are on a small scale, it may be relatively easy for him not only to suppress them temporarily, but also, by that action, to prevent their resuscitation on either that or larger scale. It has been shown countless times in history that where firm enemy action has been taken in time against small beginnings, such action has always met with success. To counter this, therefore, it is again important that the commencement of guerilla operations should be on the highest and widest scale that the area concerned will permit.

    The two arguments above overwhelmingly support this policy.

G.S.(R)

(b) The second point to be noted is that the organization of guerillas must not be of a higher degree than circumstances will, with reasonable safety, and a view to efficiency, permit.

The factor of "safety" concerns possible enemy counter-action; the closer and higher the organization, the more easily can it be broken up and become ineffective. It is valueless and dangerous prematurely to organize partisan bands, acting independently as they normally should, into platoons, companies, squadrons, etc. and then into regiments or brigades, with nominated commanders, skeleton orders of battle, intelligence services, etc.; such organisation necessitates documents, written orders, files, etc., all or any of which, falling into the enemy's hands, may enable him to destroy the guerilla movement at a blow.

In any case, such organization is unnecessary in the early stages. In these conditions, except for a central directing brain and a few trusted emissaries, partisan bands should be self contained, acting under their own leader's initiative towards the ends directed by the. controlling authority, obtaining their own information by the most direct and simplest means (usually by word of mouth) and maintaining the loosest organisation compatible with effective action.

6. The factor of efficiency concerns the inherent advantages that guerillas enjoy through their superior mobility and their lack of communications. A premature tightening of organisation is directly inimical to these two advantages, so that an increase in the degree of organisation over the bare minimum necessary must inevitably lead to decreased efficiency. It is obvious, however, that, in the culminating stages of guerilla warfare, with large masses of guerillas taking an open part, some degree of organization is necessary in order to establish a chain of command, to render administrative arrangements possible, and to collect intelligence as a basis for plans, etc.

7. At any time, therefore, the correct degree of organization to be established must be a matter for the most serious consideration of the controlling authority; as conditions change, so will the degree. To meet changing circumstances, therefore, the controlling authority must plan in advance, so that closer organization can be instituted when the moment demands, or can be relaxed if enemy action temporarily necessitates.

THE ART OF GUERILLA WARFARE

Objectives of Guerilla Warfare.

8.   The whole art of guerilla warfare lies in striking the enemy where he least expects it, and yet where he is most vulnerable: this will produce the greatest effect in inducing, and even compelling, him to use up large numbers of troops in guarding against such blows.

Modern large sized armies, entirely dependent as they are on the regular delivery of supplies, munitions, petrol, etc for their operations, present a particularly favourable opportunity For guerilla warfare, directed against their communications by road, rail or water, and against their system of internal postal and telegraph communications, at focal points, which offer most suitable targets for guerilla action.

The guarding of these communications and dumps against attack will, even before the threat is evident, necessitate the institution by the enemy of detachments and posts, more particularly at vital points on the communications and where dumps of importance are located. These detachments themselves are a suitable object of attack.

Thus the operations of guerillas will usually be directed against the flanks of armies, against their communications and against posts and detachments established by the enemy for the express purpose of protecting his important localities against such sporadic attempts.

Methods and Principles.

9.   The methods and principles of guerilla warfare must be based on a proper estimation of the relative advantages and disadvantages enjoyed by the enemy on one hand, and the guerillas on the other, in armaments, mobility, numbers, information, morale, training, etc.

10. The enemy will almost invariably possess armament superior both in quantity and quality —-i.e., he will have artillery, mortars, gas, armoured vehicles, etc., in addition to the automatics and rifles with which the guerillas will also be armed. In total strength the enemy will normally have the superiority as well, but the distribution of his forces will necessitate the use of detachments against which superior guerilla forces can be brought.

11. It is in mobility, in information, and in morale that the guerillas can secure the advantage, and those factors are the means by which the enemy's superior armament and numbers can best be combated. The superior mobility, however, is not absolute, but relative - i.e. to the type of country in which the activities are

G.S.(R)

staged, to the detailed knowledge of that country by the guerillas, etc. In absolute mobility, the enemy must always have the advantage — i.e., the use of railway systems, the possession of large numbers of motors, lorries, armoured cars, tanks, etc., of large forces of cavalry, etc. By the judicious selection of ground, however, and by moves in darkness to secure surprise, the guerillas can enjoy relatively superior mobility for the period necessary for each operation.

12. The enemy will usually be in a country where the population is largely hostile, so that the people will actively co-operate in providing information for the guerillas and withholding it from the enemy. The proper encouragement of this natural situation and the development of the system of obtaining information will ensure that the guerillas are kept au fait with the enemy's movements and intentions, whereas their own are hidden from him

13. Morale, training, etc., are factors of importance in which first one side and then the other may have the advantage. Where the enemy is constrained by demands on his forces to use reserve and second-line units for guarding, communications etc., neither the morale nor training will be of a high standard. The morale of the guerilla should always he high; fighting in his own, country, among his own people, against a foreign foe who has invaded his land, the justice of his cause will inflame his embitterment. At the same time, the narrow limits of the training he requires, his natural dash and courage, and the careful, detailed rehearsal of projected coups should enable him, with the advantage of the initiative, to match even the best trained troops.

14. Guerillas must obtain and make every effort to retain the initiative. To have the initiative confers the invaluable advantage of selecting the place of operations that most favour success as regards locality, ground, time, relative strengths, etc. The initiative can, always be secured by remaining completely quiescent until the moment for the commencement of guerilla activities arrives, and then suddenly launching out against an unsuspecting enemy. To retain the initiative conferring these advantages demands a ceaseless activity, so that the enemy is prevented from getting in his blow by the constantly recurring necessity of parrying those aimed at him.

15. Until the final and culminating stages of partisan warfare where large bodies of guerillas are co-operating with the regular forces, it must be the object of partisans to avoid prolonged engagements with their opponents, unless in such overwhelming strength that success can he assured before the arrival of reinforcements. The object must be to strike hard and disappear before the enemy can recover and strike back. Therefore, the action of all partisan bands must be governed by the necessity of a secure line of retirement, for use when the moment for calling off the action arrives. It must be borne in mind too, that the immunity of

THE ART OF GUERILLA WARFARE

partisans from enemy action is a most valuable moral factor; to inflict damage and death on the enemy and to escape scot-free has an irritant and depressing effect on the enemy's spirit, and a correspondingly encouraging effect on the morale, not only of the guerillas but of the local inhabitants, a matter of considerable moment; in this sphere of action nothing succeeds like success.

16. From the above review of the circumstances of guerilla warfare. the aim of the guerillas must be to develop their inherent, advantages so as to nullify those of the enemy. The principles of this type of warfare are therefore:—

- (a) Surprise first and foremost, by finding out the enemy's plans and concealing your own intentions and movements.
- (b) Never undertake an operation unless certain of success owing to careful planning and good information. Break off the action when it becomes too risky to continue.
- (c) Ensure that a secure line of retreat is always available.
- (d) Choose areas and localities for action where your mobility will be superior to that of the enemy, owing to better knowledge of the country, lighter equipment, etc.
- (e) Confine all movements at much as possible to the hours of darkness.
- (f) Never engage in a pitched battle unless in overwhelming strength and thus sure of success.
- (g) Avoid being pinned down in a battle by the enemy's superior forces or armament; break off the action before such a situation can develop.
- (h) Retain the initiative at all costs by redoubling activities when the enemy commences counter-measures.
- (i) When the time for action comes, act with the greatest boldness and audacity. The partisan's motto is "Valiant yet vigilant"

Those are the nine points of the guerilla's creed.

Organization.

17. In guerilla warfare it is the personality of the leader that counts: It is he who has to make decisions on his own responsibility and lead his men in each enterprise. He must therefore be decisive and resourceful, bold in action and cool in council, of great mental and physical endurance, and of strong personality. These qualities alone will enable him to control his followers and win their unquestioning obedience without the close constraints of military organisation and discipline which are the antithesis of guerilla action and a drag on its efficiency.

G.S.(R)

A background of military training is invaluable for a guerilla leader, tempering his judgements and strengthening hie decisions. The almost universal adoption of compulsory military training throughout Europe and the levees en masse of the Great War will usually ensure that every leader will have had a military experience of one sort or other. To this should be added, by study and instruction, a realisation of the influences of mechanisation on the operations of large armies, both as a factor limiting and handicapping initiative and as a factor opening up new possibilities of mobility, of air action, of five power, etc.

The selection of suitable leaders is therefore of paramount importance. The central authority must, and perforce will be, some man of prestige and weight who has been a leading personality in the territory in time of peace, as the leader either of some powerful association or league or minority. Leaders of local partisan bands will be selected from those of standing or mark in the locality who possess the necessary attributes of personality.

18. It may, however, frequently be advantageous to appoint certain serving army officers for duty with guerilas, either to serve directly as commanders, more particularly in the higher spheres, or as specially qualified staff officers or assistants to guerila commanders. In such cases, it will often happen that the serving officer works hand and glove with the titular leader, the latter, owing to his local connections etc., ensuring the cohesion of his guerillas, while the former supplies to the partnership the technical knowledge necessary for the most effective direction and co-ordination of the guerilla's operations.

19. The wider the guerila movement spreads, and the closer that its organisation must ultimately in that case become, the greater will be the need for a leaven of regular officers to carry out the basic work of simple staff duties, and to affect liaison with the regular forces. These officers must, however, clear their minds of all pre-conceived ideas regarding military procedure and apply their minds entirely and objectively to the success of the matter in hand. The very fact of their being regular officers may prejudice their position in the eyes of the partisans, and such prejudice can only be overcome by the. proof they can give of their value to the guerila cause.

20. In cases where the guerillas are a nation in arms, or part thereof, fighting for their freedom in alliance with or assisted and instigated by a third power which is willing and anxious to render all assistance to them, it will usually be advisable for that power to be represented by a mission at the headquarters of the guerila movement. The duties of such a mission would be to provide expert advice, to ensure liaison, to arrange the supply of arms, ammunition, money, etc., and to provide leaders and assistants to leaders, if such were found to be necessary.

21. It is of great importance that the personnel of such missions should he au courant with the countries and territories where they are to

work: the more detailed knowledge, personal liaison and reconnaissance that they have or can effect before operations are even envisaged, the greater is the chance of their success. They must study the languages, dialects, topography, etc.; they must know the ethnological, political and religious groupings of the people, the history and aspirations of the country, its heroes of the present and martyrs of the past. They must in fact be prepared, at the risk of future regrets and disillusion, to identify themselves in every way with the peoples they are to serve.

22. As described in paragraph 5 (b) it s important that the degree of internal organization of the guerilas should he suited to the conditions in which they are operating; over-organization is more dangerous and detrimental to guerilla operations than too loose an organisation. The latter can be tightened as circumstances prescribe, whereas the relaxing of control that has once been established, even though necessitated by changed conditions, must at first lead to some embarrassment, confusion, and loss of direction.

23. The organization of partisans must usually commence with the formation of local bands, numbering not more than about 30 men each. It is not only simpler and more convenient to form them on a local basis, but also quicker.

The men live in the neighbourhood, they know the country, they know each other, and their leaders, and can assemble rapidly when required, either for operations in their own area, if targets for attack exist, or for transfer to some area where conditions are more favourable. At the same time, there will be many areas where it will not be possible to form bands. Suitable and willing men in such areas must be given a rallying place to which they will move under their own arrangements and there join existing bands.

24. Modern developments, particularly in aircraft, mechanized forces and wireless, have profound influences on guerilla warfare, enabling the enemy rapidly to concentrate in opposition to any moves of guerillas that have been discovered. Concealment from aircraft, therefore, becomes one of the most important factors and inevitably curtails the possibilities of large forces of guerillas moving at will throughout the country. In effect, such large forces, if they are to remain undiscovered can only move by night and must conceal themselves by day or else move by routes —i.e. through thick forests etc. —which afford concealment from reconnoitring aircraft: such routes however themselves offer some difficulty to movement.

25. In addition, areas which offer good opportunities for concealment are usually just those areas where the maintenance and supply of large guerilla forces becomes difficult. They are usually wild, with little cultivation or pasture land for carrying stock

G.S.(R)

or feeding the guerillas' animals, apd supplies would have to be brought in specially. At once the guerillas would begin to be dependent on communications, a situation cramping their mobility and exactly opposed to the characteristic which constitutes their chief military value.

26. It must be clearly realised therefore that in most European countries, except for large areas in the east and south-east conditions will rarely at the commencement of a campaign be suitable for the employment of guerillas in large masses. Even in Asiatic and North African countries, the presence of hostile aircraft will make this difficult.

27. It is therefore probable that in the early stages of a war, the scale of guerilla warfare will not exceed the activities of partisan bands; even if it should never exceed this, however, a guerilla campaign of this type directed with skill and executed with audacity and ceaseless activity will be a most potent factor in absorbing hostile forces and thus rendering a proper campaign by the enemy impossible.

For this type of guerilla war a loose organization is essential, and co-ordination and direction of effort must emanate in considerable detail from the central controlling authority known as "The Chief".

"The Chief", or Military Mission or Guerilla Bureau.

28. "The Chief" may he either an individual of the country concerned located with his small staff in the area of guerilla activities, or a section of the General Staff (Intelligence Branch) of the Army concerned, and located at its General Headquarters, or even a military mission from a third party, located either at the General Headquarters of one of the armies in the field, or some other more suitable place. "The Chief" may thus be established in either friendly territory, or in territory occupied nominally by the enemy. The relative advantages of either course are as follows :-

29. If located in enemy territory —i.e. in the area where guerilla bands are to operate —contact and direction are easier, co-ordination of plans simplified, and "The Chief"'s" presence must have a stimulating effect on the partisans. In addition, intelligence and planning, which depend so much on local conditions at the moment, can be more thorough. On the other hand, the nearness of the enemy and his activities will necessitate constant changes of location, and the possibility of enemy raids will necessitate the reduction of documents, files, etc. to a minimum which may be incompatible with effective action.

30. Conversely, the installation of "The Chief" at the General Headquarters of an army, or even in friendly territory, brings in its train closer relations with the regular forces, wider sources of information, the possibility of complete documentation, greater security, and facilitates the provision of such supplies as the guerillas may receive, i.e. arms, ammunition etc. What is lost, however, is the close touch with the active agents of the guerilla campaign, and the inspiration which only the presence of "The Chief" in their midst can really arouse. This can however be counteracted by the appointment of a "Deputy Chief" specially chosen for his personality and characteristics, and granted plenipotentiary powers for use in emergency.

31. "The Chief" will direct his bands by emissaries or personal visits and will appoint regional assistant-chiefs to assist him. When a large operation is planned, he will frequently direct and lead it in person. As, however, the organization is purposely loose, it is important that "The Chief" should not be exposed to unnecessary danger. Much of his plans and intentions for future action, his knowledge of the country and of his assistant-chiefs will not have been committed to paper nor can be, but are stored in his brain; his loss might be irreparable.

32. Assistant-chiefs may again appoint sub-chiefs under them, according to the size of the regions for which they are responsible and the number of bands they contain.

Arms and Equipment.

33. The provision and replenishment of arms and equipment for guerillas is a problem that requires constant consideration. It is obvious that, if adequate supplies can be obtained before hostilities commence and can be suitably distributed, the problem is immensely simplified; further, guerilla operations can then be commenced without delay. The possibility of providing such peace stocks is governed almost entirely by political considerations, so that each country or district must be considered as a separate case; the altitude of the General Staff concerned is also of importance, more particularly in view of the pressure they can exert on their governments, a pressure which grows in weight on the approach of crises.

34. The arms most suitable for guerillas are those which do not hamper their mobility, but which are effective at close quarters. Guerilla actions will usually take place at point blank range as the result of an ambush or raid, with the object of inflicting the maximum amount of damage in a short time and then getting away. What is important therefore is a heavy volume of fire developed

immediately, with the object of causing as many casualties and consequent confusion as possible at the outset of the action.

Undoubtedly, the most effective weapon for the guerila is the sub-machine gun which can be fired either from a rest or from the shoulder i.e. a tommy-gun or gangster gun, in addition, this gun has the qualities of being short and comparatively light. Special efforts must therefore be made to equip each band with a percentage of these guns.

Carbines are suitable, being shorter and lighter than rifles, and the long range of the rifle is not necessary. After carbines come revolvers and pistols for night work and for very close quarters, and then rifles. The more silencers that can be obtained for these weapons the better. A "Silenced" rifle or pistol not only impedes detection, but has a considerable moral effect on the sniping of sentries, etc. Telescopic sights are invaluable for snipers.

Bayonets are quite unsuitable for guerillas: these are only for use in shock action which should be eschewed; a dagger is much more effective, and more easily concealed.

Bombs and devices of various kinds are of great use; when possible they should he specially made for the peculiar requirement of guerila warfare, but standard army equipment must frequently be made to serve.

35. Replenishment of stocks during a campaign, particularly of ammunition, must he a constant concern to all partisans. When operating behind the enemy's lines, the maintenance of supplies from outside will be a matter of the very greatest difficulty, frequently impossible; it is most important therefore that every opportunity to seize arms and ammunition from the enemy should be grasped. This is the only sure way of obtaining requirements. It will sometimes be necessary to organize raids whose primary object is the seizure of arms; every partisan must always have this matter uppermost in his mind, and be prepared to grasp any opportunity that offers.

Information and Intelligence Service.

36. In their normally superior facilities for obtaining information guerillas have a factor in their favour of which the fullest advantage must be taken in order to counteract the enemy's superior armament and equipment. Operating as they usually will be among a friendly populace, a system of obtaining information must be so built up that, from the offensive aspect, the fullest information required can be obtained prior to any contemplated operation : and, from the defensive aspect, no action which the enemy intends against the guerillas will escape prior detection. Further, information must always be sought giving details of the enemy's moves, detachments, convoys, etc., which may lead to the initiation of a successful operation.

THE ART OF GUERILLA WARFARE

37.   An enemy in occupation of territory is compelled to mix in varying degrees with the inhabitants. Troops must be billeted in houses: cafes and beerhouses will be used for their recreation, working parties will be employed for unloading trains, repairing roads, etc. These circumstances are extremely favourable for the collection of information by the local populace acting as agents. In fact, every reliable man, woman and child of common-sense and reliability should be encouraged and trained to keep his ears open for items of information, and where conditions are suitable, to seek for it by questions, by purloining letters, etc. Among the most suitably placed to act as agents are barbers, waitresses, domestic servants, priests, doctors, telephone and telegraph operators, postmen and camp followers generally.

38.   The collection and collation of this information requires some consideration. As pointed out earlier, the seizure of documents by the enemy from guerillas as the result of raids, interception of letters, etc., is of the greatest value to him in his efforts to crush the guerila warfare.   Messages passed by agents therefore should be verbal as far as possible, and the degree of documentation by local partisan lenders must not exceed that which allows reasonable security. As and when the guerilla organization grows tighter and closer, collation and recording of intelligence will increase until the stage is reached that at the headquarters of the guerilla forces in the field there is a proper intelligence staff with files, maps, enemy order of battle, etc. To err on the side of over-organization, however, is to court disaster; hence the overriding importance of the personality of the leaders. The leader alone, by his activity, his drive, his flair for guerilla warfare, his intelligence and wit, directs his men to successful action without the close organisation necessary for regular forces.

39.   When guerilla operations commence on whatever scale, the enemy will institute counter-measures, of which one important aspect will be intelligence. But he will be working usually amidst a hostile populace : without their co-operation his task will be more difficult and will require a larger number of his own men to carry it out.

40.   The guerillas must therefore impress on the people the vital necessity of withholding from the enemy all information about them however harmless it may seem : the people must be convinced that their refusal to co-operate with the enemy in this respect is of the greatest importance for the redemption of their country from the enemy's grasp, and for the safety of their friends and relatives. They must be warned never to discuss the activities of the guerillas in any circumstances whatever.

In every community will be found certain individuals so debased that for greed of gain they will sell even their own countrymen. Against this contingency close watch must be set, and

G.S.(R)

wherever proof is obtained of such perfidy, the traitor must be killed without hesitation or delay. By such justifiably ruthless action others who might be tempted to follow suit will be finally deterred.

41. It will be necessary, in addition, to harass the enemy's intelligence service in every possible way. Agents that he may have imported must be tracked down and shot, his intelligence officers and staffs sought out and neutralized, and captured documents and plans destroyed after perusal.

42. Guerillas themselves must be trained to give away no information if captured. The enemy intelligence officers will be adepts in leading prisoners into indiscretions, in installing listening sets and "pigeons" in prisons, concentration camps, reading prisoners' ingoing and outgoing mails, etc.

43. The advantage of superior information is the guerillas' greatest asset : it must be used to the fullest extent possible.

Intercommunication.

44. All means of communication that are open to interception by the enemy must be used with the greatest discretion —i.e. civil postal service, telephone and telegraphs, etc., as any code and ciphers used by guerillas must of necessity be simple or only infrequently changed, and their solution by the enemy will not be a difficult task. Such devices therefore only give a very relative security.

45. The passing of information verbally and direct is clearly the safest and in many ways the most reliable means. At the same time, however, opportunities for this will not always occur, and frequently messages must be written and conveyed by several hands before reaching their destination. For this purpose it is often better to use women and children who are less suspect and probably enjoy greater immunity from search

46. It will be incumbent on leaders within their own areas to arrange adequate means for the collection of information, and their own ingenuity will produce many devices, such as messages left in clefts of trees, in stone walls, in culverts, etc. Pigeons are occasionally useful, but their limitations are obvious — i.e. ease of detection, uncertainty, etc., and the greatest care must be observed in their use.

47. For messages of operational importance between partisan bands and the scouts, and within groups of partisan bands, etc., wireless offers great possibilities. It can be used by scouts to

inform their band that an enemy convoy is leaving by a certain route, offering a chance of ambush; it can be used within groups to co ordinate attacks, to pass on information, etc. The smaller the transmitting set and the wider its range the more useful it becomes ; ease of concealment is a very important factor.

Wireless should not be used except for matters of importance ; sets are not easily replaced if discovered and should be guarded preciously. It may be advisable to fix certain hours only during which wireless may be used. ALL MESSAGES IN WIRELESS MUST BE IN CODE OR CIPHER.

Training.

48. Training in the full military sense is not applicable to guerillas, but on the other hand any guerilla who has a background of military training is ipso facto a better partisan. The object of military training is to make any recruit of whatever calibre into a reasonably good soldier, so that it is based on the lowest common denominator. Guerillas on the contrary will usually be recruited from men who have a natural aptitude or a fondness for fighting, who are accustomed to the use of weapons, to hard sleep, to movement in the dark, etc.

Their training, therefore, should first be directed to the use of their basic weapons i.e., automatic rifles, carbines, pistols, rifle, etc. and to the use of the various destructive devices such as bombs, road and rail mines, etc. which are such a special and useful feature of guerilla warfare.

49. For these devices knowledge of electrical equipment is of great value; leaders must therefore endeavour to include in their bands a few men with this experience : if they do not exist, suitable men must be trained. The actual placing of these devices, and even their firing, can often be carried out in emergency by untrained personnel, but the risks of inefficacy and failure are great and should not be run for want of a little time spent in training.

50. Localities for training must be carefully selected so that surprise is impossible : it is essential to post sentries far out where enemy movement can be seen in time.

51. Weapon training of guerillas must be efficient, not only so that the men may have confidence in their weapons and shoot to kill, but also in order to save ammunition which is frequently an important factor in guerilla warfare. A few rounds spent on perfecting shooting, and testing of rifles, will be amply repaid.

52. Training in defensive action against modern weapons is of importance, more particularly in the following aspects:—

G.S.(R)

(a) <u>Aircraft:</u>
Partisan leaders must impress on all their men that the surest way of attaining success in their operations is by remaining undetected, and that detection will always be followed by enemy action against them. Concealment from aircraft is of the greatest importance, and men must be trained to take cover quickly, to lie face downwards, and to remain absolutely still until the aeroplane has passed.

(b) <u>Tanks, Armoured Cars, etc.:</u>
These are very blind when forced by fire to close down their screens; both are very susceptible to ground fire.

(c) <u>Machine Guns, etc.</u>
Smoke screens formed by smoke bombs are the best antidote.
For further details, see the Partisan Leader's Handbook.

Enemy Counter Action.

53. The first effect on the enemy of the institution of guerilla warfare will be to compel him to strengthen all posts, guards, detachments etc., and to carry out all movements in convoy, even if only of a routine nature. By this the guerillas will have achieved a part of their object, i.e. more enemy troops will be absorbed in purely protective duties, and his forces for offensive action correspondingly reduced.

This reaction of the enemy is however purely defensive. As the scale of guerilla warfare increases, and as successful attacks are carried out against those strengthened posts, convoys, etc., the enemy will undertake active offensive measures against the partisans with the object of finally crushing them.

54. Until the first, stage has been reached, and this will not be long, i.e. moving in convoy, etc., members of partisan bands may well be able to remain living undetected in their own homes, and collecting by summons for particular operations. This however will soon be rendered impossible by the searches, raids, etc., and issue of curfew, passport and other regulations that the enemy will introduce. When that moment comes it will be necessary for the partisans to "go on the run" i.e. to live as a band in some suitable area where the nature of the country enables them to be relatively secure.

55. The commencement of offensive action by the enemy will be marked by the institution of "flying columns" — detachments of from fifty to two or three hundred strong, mobile by means of horses, lorries, etc., and equipped with several days of supplies — which will be sent out to search the country, moving by circuitous

THE ART OF GUERILLA WARFARE

and haphazard routes, employing scouts and advance guards, and probably assisted by aircraft. The final stage, when this action is insufficient, will be the organization of "drives", in which large forces of troops consisting of all arms will be used to sweep through successive selected areas, and the accompanying intelligence officers, their staffs, informers, agents, etc., will interrogate every man falling into the net and arrest any to whom suspicion attaches. Aeroplanes are certain to co-operate.

56. Against flying columns, the guerillas' superior sources of information, knowledge of the country and individual mobility should be adequate protection : the object of the guerillas in those circumstances is to avoid discovery and not take military action against the flying columns unless overwhelming strength against any particular column can be combined with favourable circumstances in which to destroy it.

57. Against large scale drives the guerillas must give way and move off to some locality where the enemy is relatively inactive. It must be remembered that in countries of any large extent the number of troops required to carry out comprehensive drives simultaneously through every area subject to guerilla warfare will usually be prohibitive. Should the enemy attempt such a policy, the object of this warfare will be even nearer to achievement, i.e., rendering the enemv incapable of carrying on an effective campaign.

58. The counters to such a policy are clear. If the enemy's drives throughout the whole area affected give no chance of eventual escape, the partisans must harry the advance as it proceeds, seek the weak spots in it, and break through back into their own country, either by infiltration, or by massing Against a weak spot and bursting through by sheer strength and force of arms. To men who know the country and can move freely in the dark there is little risk of failure.

59. Against the various weapons that the enemy may employ, endowed as he will be with superior equipment of war. i.e. aeroplanes, tanks, armoured cars, etc., instructions are contained in the *Partisan Leader's Handbook*.

Of all these means, the most dangerous to the partisans is the aeroplane : they must be taught always to move and take up their positions by night, to take immediate cover from aircraft of all descriptions, and never to open fire on them unless the aeroplanes themselves attack.

60. Against action by the enemy, other than of a military nature, every step must be taken to render it inoperative. Such action will

G.S.(R)

include the institution of curfew hours, of a system of visas and cartes d'identite, of traffic regulations, of restriction on the use of motor transport, etc. In this field, it is the civilian population which can most assist the guerilla : a policy of absolute non-co-operation leavened with enlightened stupidity will do much to render the enemy's control ineffective.

Planning and Action.

61. Just as in time of peace the study of the employment of its regular forces in the event of possible wars is one of the main problems of a country's General Staff, so must the employment of guerilla forces and tactics in aid of the regular army be the object of equally close examination. Probable theatres of war and possible allies in various contingencies will lead this examination over a very wide field. Cases requiring particular study will be those in which either the home country or an ally must envisage in view of the enemy's greater strength, more complete preparation, or more rapid mobilization, against a successful invasion of its territory in the early stages of the campaign, even if only to a limited depth.

62. The object of such study is to determine the possibilities of guerilla warfare on the flanks of, but more particularly behind, the advancing hostile armies, and to make the necessary arrangements IN PEACE before the emergency arises. To delay study and preparation until a war has broken out will make the institution of a proper guerilla campaign infinitely more difficult, and in face of a strong and ruthless enemy, in all probability impossible.

The arrangements to be made must include:—

(a) The nomination of local partisan leaders.

(b) The provision of arms, ammunition, destructive devices, wireless sets, etc., and their concealment.

(c) Selection of "The Chief" and of the personnel of his staff.

(d) Provision, of ensuring liaison between General Head quarters in the field and "The Chief" with his guerillas. N.B. If "The Chief " is at General Headquarters, liaison is required between him and the deputy chief.

(e) The formulation of a plan of campaign.

(f) The selection of vital points for destruction after hostile occupation, and their preparation to that end. ETC., ETC.

63. It may well be that, among a group of two or more allied powers, one power by its wealth, its strategic position, its military experience or its initiative, is in a position to encourage and assist

the others in these preparations. Such assistance may take the following forms :—

- (a) The provision, of special weapons and destructive devices for use by guerillas.
- (b) The provision of technical experts in destructive devices specially trained to assist the leaders of partisan bands
- (c) The establishment of a mission or bureau either at the allied General Headquarters; or in the field with the guerillas, to direct operations tn co-ordination with that General Headquarters, and to arrange for the further supply and distribution of money, arms, etc.
- (d) The provision of military experts in the field to assist and co-ordinate the activities of assistant leaders.

Preparatory Planning.

64. A complete survey of likely territories must be made with a view to determining for what types of guerilla activities they will initially be suitable.

Politically, the field of action for guerilla warfare may be broadly divided into three distinct spheres:—

- (a) Where the population, except for numerically insignificant minorities, supports the hostile power. This territory usually comprises the enemy's home country and that of his allied and associate powers,
- (b) Where the population is, in varying degrees, hostile to the power in occupation.
- (c) Neutral countries.

Friendly Population.

65. Unless a war has begun in opposition to the general weight of public opinion, the enemy's home country will at the outset have been brought to a high pitch of patriotism and jingoism. Such conditions offer no scope for the organization of armed intervention by guerillas, and this type of warfare must therefore be limited to subterranean attacks by disaffected individuals or small groups against targets that will interrupt communications, interfere with or damage supplies of food, munitions, etc., assist in diverting the enemy's armed forces and generally lower the morale of the people.

G.S.(R)

66. At the same time the people's will to war must be sapped and undermined in every other way, so as to induce a craving for peace and for a change in the regime of the country which will lend to it. The object must be to prepare a situation in which an increasing and vocal part of the population will be opposed to the government and its policy, and any alternative will seem to offer fairer prospects. At the right moment it will be desirable to focus public opinion on to an alternative leader or party.

67. Such a campaign is best carried on by "whispering" by skilful propaganda through the press and wireless, by magnification of hardships such as food restrictions, by the sabotaging of food supplies, communications, by publishing exaggerated casualty lists, etc., and many other means. Even in the final stages of such a campaign, however, there is no field for the employment of partisan bands; there representatives either of a foreign power or a disaffected minority, would only serve to exacerbate the patriotism of the general population. What is required is to divide the population of the enemy against itself : the means are endless — knowledge of the country and a fertile imagination will devise the methods.

Hostile Population.

68. A population hostile to the enemy's occupation offers immediately a sphere for the fullest development of guerilla warfare in all its aspects, culminating in a general rising of the people against the enemy. The types of warfare to be employed at the outset must depend on the nature of the country; it is clear that in highly cultivated districts with few physical features the concentration of partisan bands into large formations is out of the question until such time as the enemy's hold begins through weakness to relax. Then is the moment for a general Levee en Masse of the population with such arms as they have concealed or seized ; the enemy's defeat will not long be delayed.

69. In cases of this nature the provision of arms and ammunition and arrangements for replenishing stocks are of primary importance. Where the possibility of aggression by a hostile power and the occupation by it of foreign territory can be foreseen, such provision should invariably be made before the commencement of hostilities. Not only can adequate stocks be more easily obtained and planted, but also more thorough precautious can be made for secrecy in delivery and in distribution and storage.

Where such provision cannot be made beforehand, an organization must be immediately created for the running of weapons and explosives from neutral or friendly countries, and plans must be

THE ART OF GUERILLA WARFARE

worked out and put in hand for the seizure of hostile stocks by local guerillas.

70. In general, the action to be undertaken in areas where the people are hostile to the occupying power is to stimulate the morale of the inhabitants, to create a policy of complete non co-operation, both active among those best fitted for it and passive by the whole of the remainder. It is necessary to convince the people that the hostile power is not de facto in control, that its writ does not run and that it will eventually be compelled to evacuate the territory, when those who have tacitly accepted its control will be punished, and those who have opposed it will be rewarded.

Neutral Countries.

71. The institution, of guerilla- activities in neutral territories from which the enemy draws supplies must depend to some extent on the political and other relations between the powers concerned. In certain cases it may be politic to ignore the assistance given to the enemy by a particular neutral country in view of the even greater aid that is being received. When however, the supplies which the enemy is obtaining are vital to his conduct of the war it may be necessary actively to hinder this provision in spite of otherwise friendly relations with the country concerned, and to risk the rupture of such relations.

72. This risk, however, must be reduced to a minimum and postponed as long as possible. Its elimination depends primarily on the skill with which the campaign is carried out. The methods, to be employed to hinder supply range from the purchase of supplies over the head of the enemy, the organization of labour strikes at the vital points —i.e., factories, mines, docks, etc. to the sabotaging of ships, trains and machinery. The engagement of local firms of solicitors, not too scrupulous and at the same time experienced in neutrality and labour legislation, and in the procrastination of judicial procedure will be of the greatest assistance.

73. As in the case of guerilla warfare proper, this a subject which requires close study and preparation before hostilities commence, and the selection of suitable personnel, experienced in shipping and commerce generally, and maritime and neutrality laws of the countries concerned.

Geographical.

74. The geographical study of a territory is concerned with two factors :

G.S.(R)

(a) Its suitability as an area for guerilla warfare. The more broken and forested it is, the more suitable will it be.

(b) The potential targets for guerilla action which it offers. These will usually be in the shape of road, rail and river communications which the enemy would have to employ for the maintenance of his armies in the field. Vulnerable points within the enemy's own territory must also be marked. The reconnaissance of territories should, whenever possible, be carried out in time of peace by selected officers who have been grounded in the principles of guerilla warfare. Their reports will be of great assistance in formulating a plan.

Organization of Bands.

75. One of the principle reasons for insisting on the advantages of peace time preparation is that, failing such arrangements, the institution of guerilla warfare BEHIND THE ENEMY'S LINES will be a matter of the utmost difficulty. The ideal at which to aim is that when the enemy invasion takes place the men who are to become the partisans should remain in their homes with their arms conveniently concealed, and allow themselves to be over-run. They will then hold themselves in readiness to commence action under their leader the moment the order is given. Where the fronts covered by the main opposing armies are wide and broken, there will be opportunity for partisan bands to penetrate the hostile lines for operations in the enemy's rear, but when the fronts are continuous, as may frequently happen, there will be no such opportunity ; without previous provision, therefore, guerilla warfare on the enemy's lines of communication, his most vulnerable and tender spot, could only be sporadic and half-hearted.

76. Most of the great powers include in their forces formations of a para-military character such as Frontier Guards, Customs Guards, Frontier Gendarmerie, and Forest Guards, etc. These organizations, dealing as they do with the prevention of smuggling, illicit crossing of frontiers, poaching, etc., contain men with an intimate knowledge of frontier districts, trained to act by night, and to be self dependent. As frontiers frequently rest on natural boundaries such as mountains, large rivers, etc., which form good areas for guerilla activities, such men will be of immense value as the nuclei of partisan bands.

77. From a consideration of the above factors it is apparent that the institution of guerilla warfare to assist the regular armies in the defeat of the enemy is a subject which must in all its aspects be

THE ART OF GUERILLA WARFARE

considered and prepared in peace to the furthest extent possible. Such planning and action should include the following:—

(a) A careful study of the territories concerned from the point of view of geography, communications, ethnology, racial and religious habits, historical associations, etc., and a decision as to possibilities.

(b) The supply and distribution of arms, ammunition, devices, pamphlets, etc., and the instruction of potential partisans in their use.

(c) The selection and training of regular army officers in the art of guerilla warfare; these would be sent to organize and lake charge of guerilla operations in their respective areas, or to act as advisers to the local leaders. Such training should include a period of residence in the territory concerned.

Conclusion.

78. The more the subject is considered the more apparent it becomes that in guerilla warfare it is the personality of the leader which counts above everything. It is he who by his personality and steadfastness must hold the loosely organized partisans together, and by his courage, audacity and high intelligence successfully direct and lead their operations.

79. These operations range over an unlimited field according to local circumstances. Large forces of guerillas can harry the flanks of an advancing or returning army, can raid his communications in force, destroying railways, burning supply dumps and capturing convoys, and then withdraw again to the security of their own lines. Small bands of partisans can live behind the enemy's lines, or filter through gaps in his front, and carry on similar activities on a smaller scale. Individual guerillas can be permanently located in the enemy's rear, where by the sniping of guards, the destruction of military vehicles, buildings, etc., they can be a running sore in his flesh, draining his vitality and hampering his action.

80. Guerillas obtain their advantage over the enemy by their greater knowledge of the country, their relatively greater mobility, and their vastly superior sources of information. Those are the factors which, when properly exploited, enable them to engage with success an enemy who is better equipped, more closely disciplined, and usually in greater strength.

81. The main objects of guerilla warfare are to inflict direct damage and loss on the enemy, to hamper his operations and movements by attacks on his communications, and to compel him

to withdraw the maximum number of troops from the main front of battle so as to weaken his offensive power. Direct action of the types envisaged will bring the desired result about. It must always be remembered that guerilla warfare is what regular armies have most to fear. When directed with skill and carried out with courage and whole-hearted endeavour, an effective campaign by the enemy becomes almost impossible.

82. Guerilla warfare is much facilitated by the co-operation of the local inhabitants, but in the face of an uncompromising hostile occupation this will only become active as the result of successful action by the guerillas. It is this alone that will awaken in the people the spirit of revolt, of audacity and of endurance, and make them foresee and assist towards the victory that will be theirs.

83. In the modern world the time has now come when aggressor nations, to gain their ends, use every device and ingenuity that their perverted wits can devise to break down the resistance of their intended victims both before and after the occupation of their territory. Given the leadership, the courage, the arms and the preparation, however, there is one thing remaining that they cannot break, and that is the spirit of the people whose territory has been over-run, a spirit expressing itself in uncompromising and steadfast resistance to defeat and in a ruthless and uncompromising warfare of partisans until the enemy is forced to cry "Halt!" and depart. In the long history of the world such deeds have been done, such causes won; and they can be won again, given opportunity.

F I N I S

www.ingramcontent.com/pod-product-compliance
Lightning Source LLC
Chambersburg PA
CBHW050253230526
45470CB00005B/2242

www.ingramcontent.com/pod-product-compliance
Lightning Source LLC
Chambersburg PA
CBHW050324220526
45465CB00005B/2128

Join the exclusive Facebook community for the latest in data, business, and side hustling:
https://www.facebook.com/DataLikeABusiness

Take our business development courses:
https://datalikeabusiness.com/side-hustle-like-a-business-introduction

Relevant Links

Continue the conversation with us:

Check out the YouTube channel so you don't miss our next upload:
https://www.youtube.com/channel/UCNgsL1PXDIt_Rvu48nwrxMw

Subscribe to the channel so you don't miss our next upload:
Https://www.youtube.com/channel/UCNgsL1PXDIt_Rvu48nwrxMw?sub_confirmation=1

Join us on DISCORD! Gather around #TheBusinessWatercooler and talk everything business & data:
https://discord.gg/GS3MJyjGwx

Appendix

Article referencing Amazon tornado

[1] https://www.reuters.com/world/us/lawsuit-against-amazon-filed-tornado-swarm-that-left-6-dead-illinois-warehouse-2022-01-17/